YOUR KNOWLEDGE HAS VALUE

- We will publish your bachelor's and master's thesis, essays and papers

- Your own eBook and book - sold worldwide in all relevant shops

- Earn money with each sale

Upload your text at www.GRIN.com
and publish for free

Bibliographic information published by the German National Library:

The German National Library lists this publication in the National Bibliography; detailed bibliographic data are available on the Internet at http://dnb.dnb.de .

This book is copyright material and must not be copied, reproduced, transferred, distributed, leased, licensed or publicly performed or used in any way except as specifically permitted in writing by the publishers, as allowed under the terms and conditions under which it was purchased or as strictly permitted by applicable copyright law. Any unauthorized distribution or use of this text may be a direct infringement of the author s and publisher s rights and those responsible may be liable in law accordingly.

Imprint:

Copyright © 2018 GRIN Verlag
Print and binding: Books on Demand GmbH, Norderstedt Germany
ISBN: 9783668730496

This book at GRIN:

https://www.grin.com/document/428448

Duli Pllana

Reflection on Brouwer's Fixed Point Theorem

GRIN Verlag

GRIN - Your knowledge has value

Since its foundation in 1998, GRIN has specialized in publishing academic texts by students, college teachers and other academics as e-book and printed book. The website www.grin.com is an ideal platform for presenting term papers, final papers, scientific essays, dissertations and specialist books.

Visit us on the internet:

http://www.grin.com/

http://www.facebook.com/grincom

http://www.twitter.com/grin_com

ON BROUWER FIXED POINT THEOREM

A. PLLANA

The Brouwer's Fixed Point Theorem [5] is one of the most well known and important existence principles in mathematics. Since, the theorem and its many equivalent formulations or extensions are powerful tools in showing the existence of solutions for many problems in pure and applied mathematics, many scholars have been studying its further extensions and applications.

The Brouwer Theorem itself gives no information about the location of fixed points. However, effective ways have been developed to calculate or approximate the fixed points. Such techniques are important in various applications including calculation of economic equilibria.

Because Brouwer Fixed Point Theorem has a significant role in mathematics, there are many generalizations and proofs of this theorem. In this paper, we will try to show several proves of Brouwer Fixed Point Theorem. First, let's take a look at Brouwer Theorem from real world illustrations. There are several real world examples, and we will take in consideration few of them.

1. Brouwer Fixed Point Theorem on $\mathbb{R}^1$

Take two sheets of paper with identical images one lying directly above the other. If you fold the top paper and change its shape completely, and you put on the top of the other sheet. The Brouwer Theorem says there is at least one point on the top paper that is directly above the corresponding point on the bottom sheet.

For instance, suppose we have two identical deck's cards. Let's keep one of the cards unchanged, and let the other rotate and stretch (fold) it. However, we do not cut or tore. Furthermore, let's take the deformed card and put on top of the unchanged card. According to Brouwer's Theorem, there must be a point on the deformed deck's card that will map exactly on the same point of the unchanged deck's card.

Another example on Brouwer's Theorem that is similar to previous examples is if we take two identical same disks. We place one on the top of the other. Then we change the shape and the form of the disk on the top, and on the other hand, we do not make any change to the bottom disk. Moreover, we put the deformed disk on the top of the disk that is on the bottom. Brouwer Theorem tells us there is at least one point of the top disk that is mapped exactly on the same point of the bottom disk.

As we mention early, we are going to present several proves of Brouwer's Theorem by starting from the simplest one on $\mathbb{R}$.

FIGURE 1.1. Mapping of folded deck's card on the unfolded deck's card

Theorem 1. *If f is continuous function and it is mapping $f : \mathbb{R} \to \mathbb{R}$ is a compact convex set in to itself there is a point x_0 such that $f(x_0) = x_0$.*[1]

Proof. The domain $[a, b]$ is mapping in codomain $[a, b]$ then $f(a) \geq a$ and $f(b) \leq b$. Let define a continuous function $g(x)$ in the closed interval $[a, b]$ such that $[a, b] \in \mathbb{R}$ where $g(x) = f(x_0) = x_0$.

By applying Intermediate Value Theorem, we can write as follows, $f(a) < g(x) < f(b)$. Then it follows $g(a) < 0$ and $g(b) > 0$ where $g(a)$ and $g(b)$ are with opposite signs. There exists a $x_0 \in [a, b]$ such that $g(x) = f(x_0) - x_0$ we will get $f(x_0) - x_0 = 0$
Hence

$$f(x_0) = x_0.$$

□

We will present two short lemmas, which plays a crucial role on implying Brouwer's Theorem. First, we will show Lemma 2 the mapping of a function to itself does not have to be a bijection in order to have fixed point. The other Lemma contributes on proving Brouwer's Theorem by retraction.

Lemma 2. *If X has fixed point property that is any continuous function, $f : X \to X$, then any space homomorphic to X has fixed point property.*

Proof. Let Y be homomorphic to X, and function $h : X \to Y$ homomorphic,

$$\begin{array}{ccc} g : Y \to & & Y \\ \uparrow h & & h^{-1} \downarrow \\ f : X \to & & X \end{array}$$

defined by $f = h^{-1}g \circ h$ is contained in X ,and X has fixed point property. There is $x_0 \in X$ with $f(x_0) = x_0$. That means $h^{-1}(g(h(x_0))) = x_0 \Rightarrow g(h(x_0)) = h(x_0)$. So, g has a fixed point. Thus Y has fixed point property. □

---

[1] There are many statement of the Brouwer's Fixed Point Theorem. But in the Theorem 1, we will take in the consideration the simplest form and prove it. If there is f a continuous function and $f : [a, b] \to [a, b]$ where $[a, b] \in \mathbb{R}$ there is a fixed point.

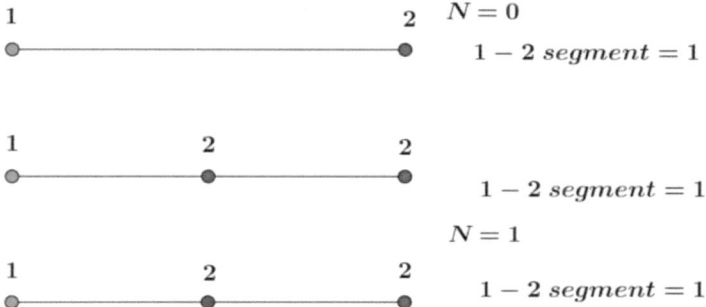

FIGURE 2.1. Induction proof for $N = 1$.

Lemma 3. *There is no retrtraction from S^1 to D^2.* [2]

2. Combinatorial: Sperner's Lemma

Emanuel Sperner is known for two theorems from 1928. Sperner's Lemma states that every Sperner coloring of a triangulation of an n - dimensional simplex contains a cell with a complete set of colors. The lemma was proven by Sperner. Later, it was noticed this Lemma implies Brouwer Fixed Point Theorem without explicit use of homology.

A Sperner's Labeling [7] is a labeling of a triangulation with the numbers 1, 2,and 3 such that

- The three corners of the triangle are labeled 1, 2, and 3.
- Every vertex on the line connected vertex i and vertex j is labeled i or j.

The simplest proof of Sperner's Lemma is by considering 1-dimensional space. Here we have a line segment (a, b) subdivide into smaller segments, and let's color vertices of subdivision with two colors. There is required that a, and b receive different colors. The number of small segments that receives two colors is odd.

Corollary 4. *If $[a, b]$ is a $1 - 2$ segment subdivided in to N subintervals labeled by Sperner's labeling, the number of the $1 - 2$ segment in the subdivision is odd.*

Proof. We show by induction the number of the $[1 - 2]$ segment is odd. First, we will check subintervals with the $1 - 2$ segment, when the number of subdivision $N = 0$ and $N = 1$

Assume that for a finite number of subdivision is N subintervals with an odd $1 - 2$ segment. If the subdivision has $N + 1$ subintervals it is going to yield an odd numbers of the $1 - 2$ segment.

---

[2]The general proof of Lemma 3 will be on Section 6.

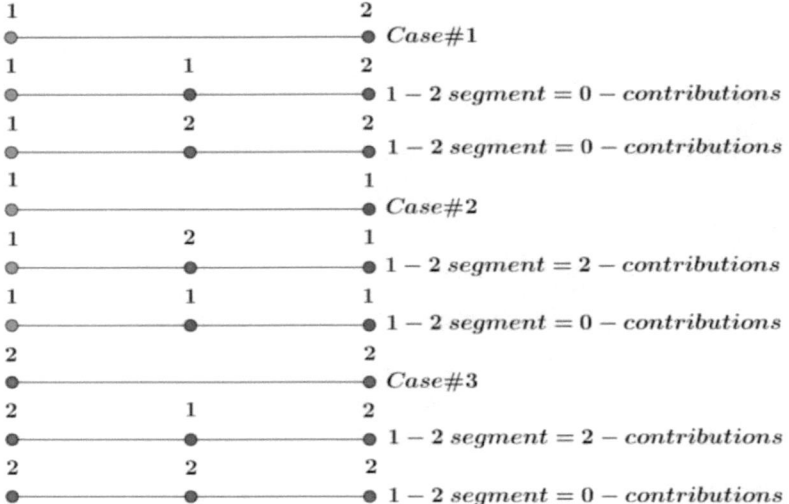

FIGURE 2.2. Three cases of the induction proof for $N+1$

If we subdivide any subinterval of the finite subdivision of N subintervals, there will appear three cases. Case #1: If the subinterval is labeled with $1-2$ segment and we subdivide into two subintervals, there will be two choices. There is a choice $1-1-2$ segment, and the other choice is $1-2-2$ segment. Both choices will contribute zero $1-2$ segment (See Figure 2). Case #2: If the subinterval is labeled with $1-1$ segment and we subdivide into two subintervals, there will be two choices. There is a choice $1-1-1$ segment that will contribute zero $1-2$ segment, and the other choice is $1-2-1$ segment will contribute with two $1-2$ segment (See Figure 2). Case #3: If the subinterval is labeled with $2-2$ segment and we subdivide into two subintervals, there will be two choices. There is a choice $2-2-2$ segment that will contribute zero $1-2$ segment, and the other is $2-1-2$ segment will contribute with two $1-2$ segment (See Figure 2). In three cases the number of contributions is even. Adding an even number and an odd number results into an odd number, and it completes the proof that $N+1$ subintervals will have an odd number of subintervals with $1-2$ segment. □

Let's start with Sperner's Lemma [8] in $Dimension-2$ by letting P be a polygon of the plane, and consider a triangulation of the polygon. Color each vertex of triangulation by one of three colors 1, 2, and 3. An edge is called *1-2 edge* if its

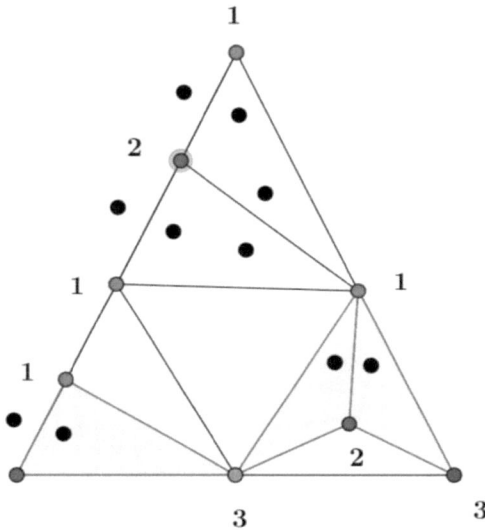

FIGURE 2.3. Complete triangles with different labels inside the polygon (Sperner's Lemma)

endpoints are colored 1, and 2 , and triangle is said to be complete if each of its vertices are colored using different colors as is shown in Figure 2.

Lemma 5 (Sperner's Lemma 1928). *If we have a polygon whose vertices are colored by three colors and triangulation is given by for this polygon, the number of complete triangles (with three different colors) is equal to the numbers of 1-2 edges on the boundary of the polygon (mod 2).*

Proof. Put a dot on each side of each *1-2 edge - segment*. We want to count the number of dots inside the triangle in two different ways. Firstly, each interior segment contributes whether 0 or 2 dots (depends weather it is a *1-2 edge*), each boundary segment contributes 0 or 1 dots. The number of dots is equal to the number of *1-2 edges* on the boundary of polygon (mod 2).

Now we count the number of dots in each triangle in the Figure 2. Complete triangles contain one dot (check this), while other triangles an even number of dots. Therefore, the number of dots is equal to the number of complete triangles (mod 2). This completes the proof. □

3. Sperner's Lemma implies Brouwer's Fixed Point Theorem

Theorem 6. *Every continuous function from a triangle to itself has a fixed point.*

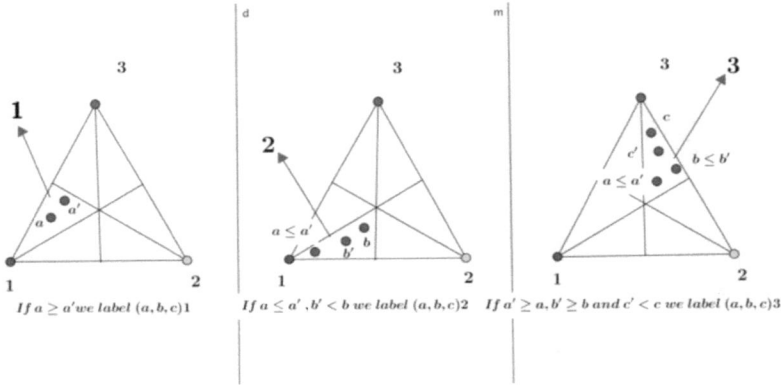

FIGURE 3.1. Representation of three cases of labeling triangles when there is NOT a fixed point

Proof. Let f be a continuous function from a triangle T in to itself. we will write $(a,b,c) \to (a',b',c')$ is $f(a,b,c) = (a',b',c')$ in Barycentric coordinates with respect to the vertices of T.

We will label each point on the triangle as follows. Suppose $(a,b,c) \to (a',b',c')$.
If $a' < a$, we label (a,b,c) 1.
If $a' \geq a$, but $b' < b$ we label (a,b,c) 2.
If $a' \geq a$ and $b' \geq b$, but $c' < c$ we label (a,b,c) 3.

If we cannot label the point (a,b,c), then $a' \geq a, b' \geq b,$ and $c' \geq c$ so $a = a', b = b', c = c'$ and the point is a fixed point. For the remainder of the proof, we assume that any point we need can be labeled; if not, it is fixed and we are done. [3]This will be formalized as follows. First, we need to show that this labeling is in fact a Sperner's labeling.

If we look at corners, we note
a) If $(1,0,0) \to (a,b,c)$, $a < 1$ unless this is a fixed point, so this corner gets label 1.
b) If $(0,1,0) \to (a,b,c)$ then $a \geq 0, b < 1$ (unless this is a fixed point), so this corner gets label 2.
c) Similarly, $(0,0,1)$ gets label 3.

If we look at the point $(a,b,0)$ on the line connecting $(1,0,0)$ and $(0,1,0)$, we see that if $(a,b,0) \to (c,d,e)$ the $c < a$ or $b < d$, so the label is 1 or 2. If not, we have $c \geq a, b \geq d$ so $c = a, b = d$ and $e = 0$ and the point is a fixed point.

Similarly, we can show that the point on the line connecting $0,1,0$ and $(0,0,1)$ get labels 2 or 3, and the points in the line connecting $(0,0,1)$ and $(1,0,0)$ gets label 3 or 1. Thus if label the triangulation in this way we get a Sperner's labeling.

We know consider a sequence of triangulations with diameter going to zero. (The diameter of a triangulation is defined to be the maximum distance between

---

[3]What we are intuitively with this labeling is picking a corner the point does not move closer to. We will eventually get a sequence of a small $1-2-3$ complete triangles converging to a point. If this was not a fixed point, the continuity of f would suggest that as our triangle approach the point, all the corners would have the same labeling.

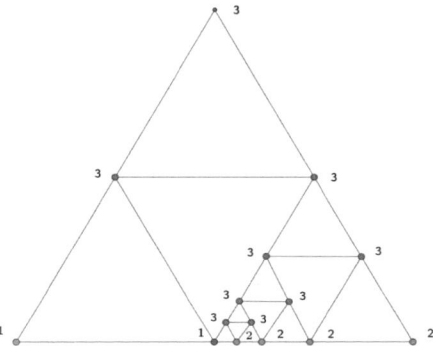

FIGURE 3.2. Convergence of labeled triangles in 2-Simplex

the adjacent vertices in a triangulation). Each of this triangulations have at least one baby $1-2-3$ triangle, suppose these triangles have vertices[4]

$$(x_n, 1, y_n, 1, z_n, 1)$$
$$(x_n, 2, y_n, 2, z_n, 2)$$
$$(x_n, 3, y_n, 3, z_n, 3)$$

with labels 1,2, and 3 respectively. Here the n indicates that the triangle is from the nth triangulation in the sequence of triangulations with diameter going to zero. We now use the Bolzano-Weierstrass to find a convergent sequence

$$(x_{nk}, i, y_{nk}, i, z_{nk}, i) \to (x'_n, y'_n, z'_n)$$

---

[4]We may construct many triangulations. For instance, we triangulate a triangle $1-2-3$ in six small triangles (see Figure 3.3 and Figure 3.4) . Then, all six triangles we subdivide in six smaller triangles. This procedure may continue endlessly. First triangulation will have at least on triangle that have all three vertices labeled with different colors. Second Triangulation (second triangulation has smaller diameter than first triangulation's diameter) has at least one triangle that have all three vertices with different colors. Third triangulation has a smaller diameter than second triangulation. Third triangulation, has at least a triangle with three vertices with different colors. This procedure goes on until we reach the smallest triangulation. According to Figure 3 and Figure 3 sequences of triangulations' diameters decrease monotonically. Considering triangulations $A_n = \frac{A_1}{6^n}$ as n goes to infinity A_n goes to zero as follows $A_1, \frac{A_1}{6}, \frac{A_1}{6^2}, ..., 0$. It is clear the triangulations' diameters are non empty sets and they are decreasing such that $A_{n+1} \subseteq A_n$. By using test of monotonic sequences test $\frac{A_{n+1}}{A_n} = \frac{\frac{A_1}{6^{n+1}}}{\frac{A_1}{6^n}} = \frac{1}{6} < 1$, the function of sequences of triangulations' diameter is monotonic strictly decreasing. The triangle $1-2-3$ is closed (bounded), and by viewing from monotonic sequences perspective there is a limit to zero. Every sequence that is bonded and has a limit it is a Cauchy sequence. The function of sequences of triangulations' diameter is a Cauchy's sequence. There is $N \in \mathbb{N}$ such that $A_N < \epsilon$. Let $s, t > N$ and $s, t \in \mathbb{Z}$ such that $d(A_s, A_t) < \epsilon$. Applying inequality yields $d(A_s, L) + d(L, A_t) < \frac{\epsilon}{2} + \frac{\epsilon}{2} = \epsilon$. Since the Cauchy sequence is convergent and the space is bounded, it is complete metric space. Moreover, $d(A_s, A_t) \leq A_N < \epsilon$. Thus, there is $x \in \cap_1^\infty A_n = 0$. On the other hand, by using the same argument in the image there is $f(x) \in \cap_1^\infty A_{N'} = 0$. Intersections of diametres are proper subsets of each other $\cap_1^\infty A_n \subseteq \cap_1^\infty A_{N'} = 0$ i.e. $A_N \to A_{N'}$ implies $f(x) \geq x$, respectively $f(x) = x$.

for $1 \leq i \leq 3$. We can do this for all sequences at once because they are the corners of triangles whose diameter tends to 0. If we wanted to avoid Bolzano-Weierstrass Theorem an alternate approach is to use sub-triangulations and find nested $1-2-3$ triangles.

Now if
$$(x_{nk}, i, y_{nk}, i, z_{nk}, i) \to (x'_{nk}, i, y'_{nk}, i, z'_{nk}, i)$$
and
$$(x, y, z) \to (x', y', z')$$
we have (by Sperner's labeling)
$$x'_{nk}, 1 \leq x_{nk}, 1$$
so by continuity
$$x' \leq x$$
Similarly, $y' \leq y$ and $z' \leq z$ so (x, y, z) is a fixed point. $\square$

Here we will present a proof of Brouwer's Theorem using Sperner's Lemma. The proof is more straight forward and direct, and it also uses familiar and simple concepts. It should be noted that while the standard proof is non-constructive, this proof both allows for more intuitive view of why the theorem is true and gives a semi-constructive approach to find the fixed point.

Theorem 7. [3] *Every continuous function from the closed unit ball in $\mathbb{R}^n$ to itself has a fixed point.*

Proof. Let T be the n-simplex in $\mathbb{R}^{n+1}$ defined by the set of points
$$\left\{ (x_1, x_2, ..., x_{n+1}) \mid x_i \geq 0, \sum_{i=1}^{n+1} x_i = 1 \right\}.$$
Let $f: T \to T$ be continuous and define
$$f(x) = f(x^1, x^2, ..., x^{n+1}) = \left(f(x)^1, f(x)^2, ..., f(x)^{n+1} \right).$$
We first make observation that for a point $x \in T$, if $f(x)^i - x^i \geq 0$, for all $1 \leq i \leq n+1$, then x must be a fixed point. This allows from the fact that both x and $f(x)$ are in T and so both points must have coordinates which sum is 1.

For point $x \in T$, define k to be any index i which minimizes the quantity $f(x)^i = x^i$. Note that k is defined for each point in T. Assign the color c_k to each vertex of T based on its value of k. Note that the jth vertex of T is the point with $x^j = 1$ and $x^i = 0$ for all $i \neq j$. Thus the jth vertex of T has the property that $f(x)^j - x^j \leq 0$ and $f(x)^i - x^i \geq 0$ for all $i \neq j$. Moreover, in the case of equality, the jth vertex is a fixed point and we are done. Assuming inequality, this vertex must be colored c_j. This shows that each vertex of T is colored with a distinct color according to its index.

Take Barycentric subdivision τ_1 of T. Color all vertices of τ_1 in the same way as vertices of T were colored. If x is a vertex in τ_1 on τ^j, then $x^j = 0$. Then $f(x)^j - x^j \geq 0$ and we see that x cannot be colored with c_j unless it is a fixed point. Thus we have a Sperner's coloring of τ_1. One simplex of τ_1, by Sperner's Lemma, has full color, so let call this simplex σ_1. Now take subdivision of τ_1 to obtain τ_2. By the same argument, labeling the vertices in the same way gives a

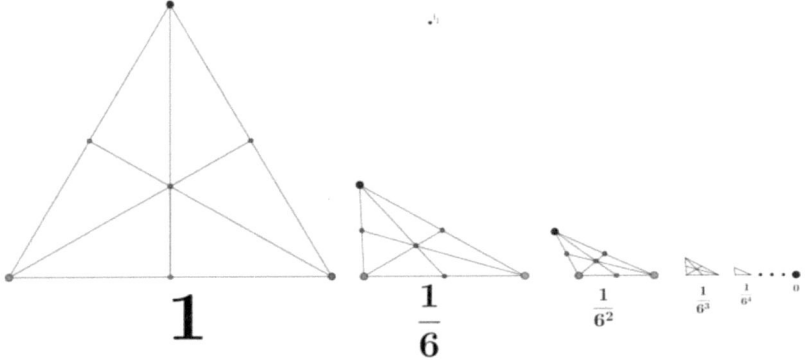

FIGURE 3.3. Convergence sequences of triangulation's diameter with barycentric subdivision that has Sperner's labeling of vertices with different colors

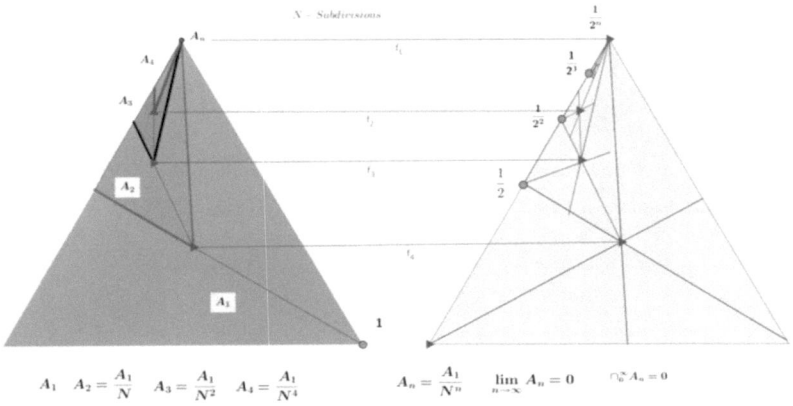

FIGURE 3.4. Convergence of Cauchy sequences in complete metric space

Sperner's coloring of τ_2 and so there exists some simplex σ_2 with full color. Continue in this way to obtain a sequence of simplexes, $T = \sigma_0, \sigma_1, \sigma_2, \ldots$

Consider the sequence of vertices from $\sigma_0, \sigma_1, \sigma_2, \ldots$ which are colored c_k. Call this sequence (p_n^k). Since (p_n^1) is bounded, it has a converging subsequence (p_{nj}^1). Now let the p be the limit of (p_{nj}^1). Note that since the diameter σ_i goes to 0 as i goes to ∞, any neighborhood of p will contain all but finitely many of the simplexes (σ_{nj}). Thus the induced subsequence of each (p_n^k) all converge [5]to p.

---

[5]p^i is a convergent sequence in X . Let

Note that because of how each sequence is defined, we have, $f\left(p_{nj}^{k}\right)^{i} \leq \left(p_{nj}^{k}\right)^{i}$ for each $i = 1, 2, ..., n+1$. Since f is continuous, we can take the limit and obtain $f(p)^{i} \leq p^{i}$. But this is only possible if $f(p) = p$. □

4. Brouwer Fixed Point Theorem for C' - map

Referring to the Brouwer's Theorem proof of [4] we will start by the Lemma 8. First, let $\mathbb{R}^{n}$ have its usual inner product $\langle \cdot, \cdot \rangle$ and let $\|\cdot\|$ be the induced norm. Let $B^{n} := \{x \in \mathbb{R}^{n} : \|x\| < 1\}$ be the open unit ball, $\bar{B}^{n} := \{x \in \mathbb{R}^{n} : \|x\| \leq 1\}$ be the closed unit ball, and $S^{n-1} := \{x : \|x\| = 1\}$ the unit sphere in $\mathbb{R}^{n}$.

Lemma 8. *There is not C^{1} map $f : \bar{B}^{n} \to S^{n-1}$ such that $f(x) = x$ for all $x \in S^{n-1}$.*

Proof. Assume toward a contradiction, where $f : \bar{B} \to S^{n-1}$ exists for $t \in [0, 1]$, let
$$(4.1) \qquad f_{t} = (1-t)x + tf(x) = x + tg(x)$$
where $g(x) = f(x) - x$. Note that for all $x \in S^{n-1}$
$$\|f_{t}(x)\| \leq (1-t)\|x\| + t\|f(x)\| \leq (1-t) + t = 1$$
therefore, $f_{t} : \bar{B}^{n} \to \bar{B}^{n}$ for all $x \in S^{n-1}$
$$f_{t}(x) = (1-t)x + tf(x) = (1-t)x + tx = x$$
Thus f_{t} fixes all points in S^{n-1}. As f is C^{1}, the same is true for g, therefore, there is a constant K such that for all $x_{1}, x_{2} \in \bar{B}^{n}$.
$$\|g(x_{2}) - g(x_{1})\| \leq K \|x_{2} - x_{1}\|.$$
Assume there are distinct points x_{1} and x_{2} in $\bar{B}^{n}$ with $f_{t}(x_{1}) = f(x_{2})$. This implies $x_{2} - x_{1} = t(g(x_{1}) - g(x_{2}))$ it follows,
$$(4.2) \qquad \|x_{2} - x_{1}\| \leq \|g(x_{1}) - g(x_{2})\| \leq \|x_{2} - x_{1}\|$$
If $x_{1} \neq x_{2}$, it implies $Kt > 1$. While $t > \frac{1}{K}$ and the function $f_{t} : \bar{B}^{n} \to \bar{B}^{n}$ is injective function. Let $G_{t} = f_{t}\left[\bar{B}^{n}\right]$ be the image of the open ball under f_{t}. The derivative of f_{t} is viewed as a linear map $f'_{t}(x) : \mathbb{R}^{n} \to \mathbb{R}^{n}$ is given by,
$$f'(x) = I + tg'(x)$$
Where I identity map on $\mathbb{R}^{n}$. As g is C^{1} there is a t_{0} such that $det f'_{t}(x) > 0$ for all $t \in [0, t_{0}]$. By inverse function theorem G_{t} is an open set for all $t \in [0, t_{0}]$. By making t_{0} small we also have f_{t} is injective for all $t \in [0, t_{0}]$.

Let $G_{t} = \bar{B}^{n}$ for all $t \in [0, t_{0}]$. Assume that is not the case. The boundary ∂G_{t} will intersect the open ball $\bar{B}^{n}$ at some point y_{0}. By the compactness we can say that there is a subsequence by the assumption that $\lim_{t \to \infty} f_{t}(x_{0}) = y_{0}$ for some $x_{0} \in \bar{B}^{n}$. Moreover, by continuity of f we have $f_{t}(x_{0}) = y_{0}$ where G_{t} is open and open sets are disjoint from their boundaries. Then y_{0} is not on $G_{t} = f\left[\bar{B}^{n}\right]$ thus

---

$$p = \lim_{i \to \infty} p^{i}$$

We will show that $p \in \cap_{i=1}^{\infty} Fi$. Let $i \in \mathbb{N}$. Let $x_{j} \in F_{j} \subseteq F_{i}$ for any $j \geq i$. Thus the sequence $(p^{i}, p^{i+1}, ...)$ is a sequence in F_{i} and is a subsequence of (p_{n}^{k}) that converges to p. This implies $p \in \bar{F}_{i} = F_{i}$. Thus $p \in \cap_{i=1}^{\infty} F_{i}$. Next we show that $\cap_{i=1}^{\infty} F_{i}$ is a singleton. To see this, let $p, q \in \cap_{i=1}^{\infty} F_{i}$ and $\varepsilon > 0$. Then there is an $N \in \mathbb{N}$ such that $diam(F_{i}) < \varepsilon$. Since $p, q \in F_{i}$, it follows that $d(p, q) \leq diam(F_{N}) < \varepsilon$. This is true for any $\epsilon > 0$. Hence, $d(p, q) = 0$, which means $p = q$.

$x_0 \in \bar{B}^n \setminus B^n = S^{n-1}$, which contradicts assumption that y_0 is in $\bar{B}^n$. For $t \in [0, t_0]$ the map $f_t \colon \bar{B}^n \to \bar{B}^n$ is a bijective function.

Let define function $F \colon [0,1] \to \mathbb{R}$ by

(4.3) $$F(t) = \int_{\bar{B}^n} det f'(x)\, dx = \int_{\bar{B}^n} det(I + tg'(x))\, dx$$

Where dx is the volume in $\mathbb{R}$, which is a polynomial in t for the function $t \in [0, t_0]$. The function $f_t \colon \bar{B}^n \to \bar{B}^n$ is bijection, and by changing the variable on multiple integrals $F(t)$ is just the volume of image $f_t[\bar{B}^n] = \bar{B}.^n$ That is,

$$F(t) = Volume(\bar{B}^n)$$

for all $t \in [0, t_0]$. The polynomial that is constant on an interval it is constant everywhere. So, $F(t) = Volume(\bar{B}^n)$ for all $t \in [0, t_0]$, and $F(1) = Volume(\bar{B}^n) > 0$. Another hand, $f_1(x) = f(x) \in S^{n-1}$ for all x, then

$$\langle f_1(x), f_1(x) \rangle = \|f_1(x)\|^2 = 1$$

for all x. Thus for any vector $v \in \mathbb{R}^n$,

$$2\langle f_1(x)v, f_1(x) \rangle = \frac{d}{dt} \langle f_1(xt + tv), f_1(x + tv) \rangle |_{t=0} = 0$$

This allows that range of $f'(x)$ is contained in $f^-(x)$ the orthogonal complement of $f(x)$. But the rank $f'(x) \leq n-1$ for all $x \in \bar{B}^n$ then $det f_1'(x=0)$ for all $x \in \bar{B}^n$. Hence,

(4.4) $$F(1) = \int_{\bar{B}^n} det f_1'(x)\, dx = 0$$

This contradicts $F(1) > 0$ and the proof is complete. $\square$

Theorem 9. *Let $f \colon \bar{B}^n \to \bar{B}^n$ be a C^1-map of unit ball in $\mathbb{R}^n$ to itself. Then f has a fixed point.*

Proof. Since fixed point property is preserved under homeomorphism we can use equivalent norm $\mathbb{R}^n$. We will use the max norm $\|x\|_\infty := \max\{|x_i| : 1 \leq i \leq n\}$. Let us write $f = (f_1, ..., f_n)$. Given $\epsilon > 0$ by Weierstrass approximation theorem, we can find a polynomial g_i such that $|f_i - g_i| < \epsilon$ for $\|x\| \leq 2$. We let $g := (g_i, ..., g_n)$ be vector polynomial function. We have,

$$\|f - g\| := \sup_{x \in \bar{B}} |f(x) - g(x)| < \epsilon$$

Since,

$$\|g(x)\| \leq \|g(x) - f(x)\| + \|f(x)\| < 1 + \epsilon$$

g maps $\bar{B}(0,1)$ to $\bar{B}(0, 1+\epsilon)$. We consider $h(x) := (1+\epsilon)^{-1} g(x)$. Then we have,

$$\left\| f(x) - \frac{g(x)}{1+\epsilon} \right\| \leq \|f(x) - g(x)\| + \epsilon \|f(x)\|$$
$$\leq \epsilon(1 + \|f\|)$$
$$\leq 2\epsilon.$$

If $f(x) \neq x$ for $x \in \bar{B}$, then by compactness there is an $\epsilon > 0$ such that $\|f(x) - x\| > \epsilon$ for all $x \in \bar{B}^n$. If we choose a polynomial function g such that

$\|f - g\| < \epsilon$ then h defined as above cannot have a fixed point. If $h(x_0) = x_0$, we have
$$\begin{aligned}\epsilon &> \|f(x_0 - g(x_0))\| \\ &= \|f(x_0 - x_0 + x_0 - g(x))\| \\ &= \|f(x_0 - x_0)\| \\ &> 0.\end{aligned}$$
a contradiction. □

5. Milnor and Rogers proof of Brouwer's Theorem

Before we get to the Milnor and Rogers proof of Brouwer's Theorem, Let's begin by defying few terms necessary to Stone-Weierstrass Theorem [2].

Definition 10. A Hausdorff space, X, is a topological space such that for any two distinct elements, x_1 and x_2, there exists two open sets, U_1 and U_2, such that $x_1 \in U_1, x_2 \in U_2$, and $U_1 \cap U_2 = \emptyset$.

$C(X, \mathbb{R})$ will be used to denote the set of all continuous functions $f\colon X \to \mathbb{R}$, where X is a compact Hausdorff space. In fact, $C(X, \mathbb{R})$ is actually a normed space, with norm defined as follows.

Definition 11. The uniform norm of a function, f, denoted by $\|f\|_u$ is given by $\|f\|_u = \sup_{x \in X} |f(x)|$.

Since X is compact and f is continuous, this supremum is always guaranteed to exist. Furthermore, for all $f, g \in C(X, \mathbb{R})$ and $x \in X$, we define $fg(x) = f(x)g(x)$, it follows that $fg \in C(X, \mathbb{R})$. Thus, we see that $C(X, \mathbb{R})$ actually forms an algebra over $\mathbb{R}$.

Definition 12. An algebra is a vector space over $\mathbb{R}$, $\mathcal{A}$, equipped with a metric and an associative bilinear product, $m\colon \mathcal{A} \times \mathcal{A} \to \mathcal{A}$. $\mathcal{A}$ will be referred to as unital if the operation m has an identity element. A sub algebra is a subset of an algebra under the same operation.

The following definitions define few terms related to algebra and sets of functions.

Definition 13. A set of functions, S, is said to separate points if, for all x, y where $x \neq y$, there exists some $f \in S$ such that $f(x) \neq f(y)$.

Theorem 14. Suppose $\mathcal{A}$ is a sub-algebra of $C(X, \mathbb{R})$ that separates points where X is a compact Hausdorff space. If there exists $x_0 \in X$ such that $f(x_0) = 0$ for all $f \in \mathcal{A}$, then $\mathcal{A}$ is dense in $\{f \in C(X, \mathbb{R}) \mid f(x_0) = 0\}$. Otherwise, $\mathcal{A}$ is dense in $C(X, \mathbb{R})$.

Now, most of preliminary definitions are out of the way and statement of the Stone Weierstrass Theorem, let's proceed with the statement and proof of Brouwer's Fixed Point Theorem.

Theorem 15. Every continuous map $f\colon \bar{B}^n \to \bar{B}^n$ has a fixed point. That is $x \in \bar{B}^n$ such that $f(x) = x$.

Proof. Let $f\colon \bar{B}^n \to \bar{B}^n$ be a continuous map. Then by Stone-Weierstrass theorem there is a C^1 function $pl\colon \bar{B}^n \to \mathbb{R}^n$ such that $\|f(x) - pl(x)\| \leq \frac{1}{l}$ for all $x \in \bar{B}^n$. Then $\|pl(x)\| \leq \|f(x)\| + \|pl(x) - f(x)\| \leq 1 + \frac{1}{l}$ (pl - means polynomial). If $hl = \left(1 + \frac{1}{l}\right)^{-1} pl$, we have that $hl\colon \bar{B}^n \to \bar{B}^n$ and $hl \to f$ uniformly.

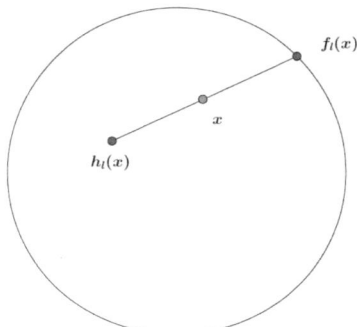

FIGURE 5.1. $f_l(x)$ - point where the ray from $h_l(x)$ to x meets S^{n-1}

Let each hl has a fixed point in $\bar{B}^n$. If not, let $f_t\colon \bar{B}^n \to S^{n-1}$ is a map. f_l - is a point where the ray from $h_l(x)$ to x meets $S.^{n-1}$ If h_l has not fixed point, this map is C^1 and has $f_l(x) = x$ for all $x \in S^{n-1}$ contradicting Lemma 8.

Let x_l be a fixed point of h_l, which is $h_l(x) = x$. As $\bar{B}^n$ is compact we can pass to a subsequence and assume that $x_l \to x_0$ for some x_0 in $\bar{B}^n$. As $h_l \to f$ uniformly this implies
$$f(x_0) = \lim_{i\to\infty} h_i(x_i) = \lim_{i\to\infty} x_i = x_0$$
Thus f has x_0 as a fixed point.

□

6. Proof of Brouwer Fixed Point Theorem by Retraction

Defining the [1] homotopy equivalence, and a relation between spaces, we first need to consider a certain relation between maps.

Definition 16. Two continuous maps $f_0, f_1\colon X \to Y$ between topological spaces are said to be homotopic if there exists a continuous map $F\colon X \times [0,1] \to Y$ (the homotopy) that F joins f_0 to f_1 i.e., if we have $F(i, \cdot) = f_i$ for $i = 1, 2$.

There [6] is a homology functor H_n for each $n \geq 0$ with the following properties: for each topological space X there is an abelian group $H_n(X)$, and for each continuous function $f\colon X \to Y$ there is a homomorphism $H_n(f)\colon H_n(X) \to H_n(Y)$, such that:

(6.1) $$H_n(g \circ f) = H_n(g) \circ H_n(f)$$

whenever the composite is defined;

(6.2) $$H_n(1_x) \text{ is the identety function on } H_n(X),$$

where 1_x is the identity function on X;

(6.3) $$H_n(D^{n+1}) = 0 \text{ for all } n \geq 1;$$

(6.4) $$H_n(S^n) \neq 0 \text{ for all } n \geq 1,$$

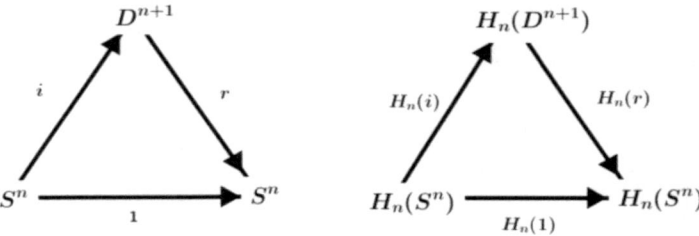

FIGURE 6.1. Diagram of abelian group and homomrphism

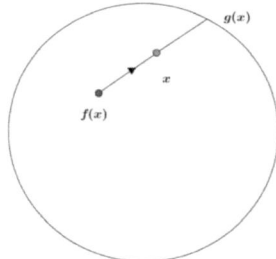

FIGURE 6.2. Fixed point on the Circle to the Disk

Lemma 17. *If $n \geq 0$, then S^n is not a retract to D^{n+1}.*

Proof. Suppose there is a retraction $r : D^{n+1} \to S^n$; then there will be a a "commutative diagram" of topological spaces and continuous maps (here commutative means that $r \circ i = 1$, the identity function on S^n).

Applying H_n gives a diagram of abelian groups and homomorphisms. By property Equation 6.1 of the homology functor H_n, The new diagram commutes: $H_n(r) \circ H_n(i) = H_n(1)$. Since $H_n(D^{n+1}) = 0$, by Equation 6.3, it follows that $H_n(1) = 0$. But $H_n(1)$ is identity on $H_n(S^n)$, by Equation 6.2. This contradicts Equation 6.4 because $H_n(S^n) \neq 0$. □

Definition 18. *A subspace X of a topological space Y is a retract of Y if there is a continuous map $r : Y \to X$ with $r(x) = x$ for all $x \in X$; such that r is called a retraction.*

Theorem 19. *Any continuous map of a closed disk in to itself (and hence of any space homeomorphic to the disc) has a fixed point.*

Proof. Suppose that $f(x) \neq x$ for all $x \in D^n$; the distinct points x and $f(x)$ thus determine a line. Define $g : D^n \to S^{n-1}$ (the boundary of D^n) as the function assining to x that point where the ray from $f(x)$ to x intersects S^{n-1} (see Figure 6). Obviously, $x \in S^{n-1}$ implies $g(x) = x$. But this is impossible, since the disk is contractible and the circle is not[6]. □

References

[1] Amanda Brouwer. The fundamental group and brouwer's fixed point theorem.
[2] Philip Gaddy. The stone-weierstrass theorem and its applications to l2 space.
[3] Kris Harper. Sperner's lemma and brouwer's fixed point theorem.
[4] Ralph Howard. *The Milinor-Rogers Proof of the Brouwer Fixed Point Theorem*. University of South Carolina Columbia USA, 2004.
[5] Sehie Park. Ninety years of the brouwer fixed point theorem. *Vietnam Journal of Mathematica*, (3):187–222, 1999.
[6] Joseph J. Rotman. *An Introduction to Algebraic Topology*. Number 119 in Graduate Texts in Mathematics. Springer-Verlag, New York, 1988.
[7] Alex Wright. Sperner's lemma and brouwer's fixed point theorem.
[8] Alex Wright. Sperner's lemma and brouwer's fixed point theorem.

---

[6]What happens when x lies on the edge of the circle? We have $r(x) = x$ or $(r(y) = y)$ when $x \in S^n$ respectively $y \in S^n$ see in figure 6.

Hence $r \circ i = id_{S^1}$ Note r is r(x). So r is actually retraction of S^n to D^{n+1}. This is contradiction by previous Lemma 3 since no retraction of the circle to the disk can exist. Therefore every continuous function mapping disk to itself has a fixed point.

YOUR KNOWLEDGE HAS VALUE

- We will publish your bachelor's and master's thesis, essays and papers

- Your own eBook and book - sold worldwide in all relevant shops

- Earn money with each sale

Upload your text at www.GRIN.com
and publish for free